ADVANCED CACAO POLLINATION TECHNIQUES

McWinner Yawman, Ph.D.

ADVANCED CACAO POLLINATION TECHNIQUES

Copyright @2021 by

McWinner Yawman

Published by

Yawman Book Publishing House

Davao City, Philippines

ISBN 978-1-7948-0916-1

ACKNOWLEDGMENTS

I would like to thank my family – Michelle, my wife, praise my daughter, Kingtor, Paul, and Attra for their continual support in this project.

Secondly, I would like to appreciate the assistance of Gladice and Therenz for your hard work and suggestions toward this project.

Finally, I am grateful to God for the health, strength, sustenance, and wisdom granted during this period.

DEDICATION

Dedicated to all cacao farmers for a double and triple harvest

TABLE OF CONTENTS

Chapter 1
THE CACAO
FLOWER

Flower stalk

The stalk or pedicel is the secondary stalk from which flowers grow off the main stem.

Sepal

The sepal is a part of the flower, usually pink for cacao. It typically functions as protection for the flower in the bud and often as support for the petals when in flower. It is found on the outermost part of the flower.[1]

Petal

Petals are also sterile floral parts that usually attract specific pollinators to the flower. All of the petals of a flower are collectively known as the corolla.[2]

Ovary

It is the female organ of a flower. The ovary contains ovules (eggs), which develop into seeds upon fertilization. The ovary itself will mature into a fruit.

Stamen

It's the pollen-producing part of a flower, usually with a slender filament supporting the anther.

Anther

It's the part of the stamen where pollen is produced.

Staminodes

They're like the stamen (the male part of the flower), but these don't produce pollen.

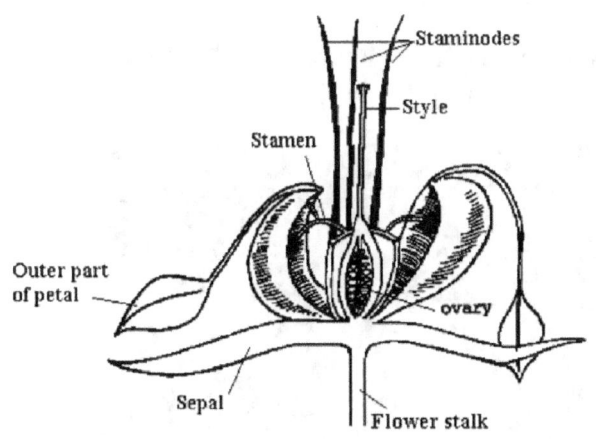

Source: Frimpong-Anin et al. (2015).

Picture of the cacao flower

Source: H. Zell Wikimedia Commons
Edited by Nancy Miorelli

3 Types of Flowers in Cacao

1. **Converging flowers**: The staminodes all point inwards at the tip.

2. **Parallel flowers:** The staminodes all
   point straight

3. **Splayed flowers:** The staminodes curve
   outwards at the tip.

The Difference between Cacao Flowers

The pollinators can successfully pollinate only the **converging** and **parallel** flower types

The cacao trees produce about **56% diverging flower** type **and 37.5% parallel type of flowers**. With these percentages, the midges are very picky and like landing on the parallel flowers the best.

This means that about 25-37.5% of the flowers are likely to be pollinated, all things being equal.

The Staminode

It's like the stamen, the male part of the flower; however, they don't produce pollen grains.

In cacao, pollination is difficult because the pollen-producing <u>anthers are hidden</u> by hoods and can only be <u>accessed by tiny insects</u> near the base of the flower.

Position of the style

The style, which is the female part of the flower, is located in the middle of all the sterile male bits (the staminodes).[3]

This means the insects must bypass the staminodes to deposit pollen on the style.

This gets more complicated because each <u>tree produces three variants of the flower.</u>

Chapter 2

FLOWERING & FRUIT DEVELOPMENT IN CACAO

Flowering

Meristematic tissues above leaf scars on the main stem and woody branches of cacao plants produce flowers.

After the second flush on jorquette branches has hardened, young cocoa plants begin to blossom. Plagiotropically budded trees, on the other hand, can flower after their second growth flush has hardened.

Typically, a single cushion can produce up to 120,000 blossoms throughout the year.

For the rest of its life, the tree's floral meristem continues to produce new flowers. From year to year, the amount of flowers produced varies dramatically.[1] The anthers discharge pollen early in the morning when the flowers open. Pollination is mostly carried out by a variety of tiny insects, including midges. As soon as anthesis occurs, unfertilized blooms fall off the plant's stalk.

Pods mature from only a small percentage of blooms (5%–10%).

Increased growth rates, earlier blossoming, and more flowers were observed after irrigation of young cocoa plants in Ghana.[2]

In the wake of drought, it was discovered that potted cocoa plants began strong flushing (and flowering) 10 days after watering each time.[3]

A comparable field study in Brazil found that water stress impeded flowering; the trees proceeded to bloom as soon as the rains started.[4]

Farm relative humidity can be raised from 50–60 percent to 90–95 percent, which induces flowering in cacao plants.[5]

Fruit development

Pods mature in about 150-200 days from the time of blossoming. An 182-day maturation period has been established.[4]

When the fruit is harvested, it is still attached to the tree. During the growth of the little fruits, there are two key points (cherelles).

To begin with, the first division of the fertilized egg occurs 40 days following flower fertilization.

The rapid growth of the fruits coincides with a significant rise in fat and starch metabolism at 75 days.[5]

Cherelles can wilt at any stage if the tree is flushed with too much water.

In contrast, few cherelles will wilt if the trees aren't subjected to a lot of flushing before pollination.

Freshly opened flowers can be pollinated synchronously by hand, resulting in healthy plants that do not wilt.

Cherelle wilt can be a fruit thinning mechanism that happens when the fruit set is not intentionally synchronized. This is due to the limited carbohydrate supply and rivalry between pods for the tree's carbohydrate reserves.

To put it another way, water stress indirectly impacts the 'wilting' process by reducing glucose uptake. A low soil level during flowering increases the amount of cherelle produced. Cacao trees' low dry-season yields can be attributed to this.

Cherelle wilt is not just a yield-limiting issue, but also indicates any carbohydrate limitation to pod growth.

Depending on the success of pollination, the number of seeds in fruit might range from 20 to 50. For the first 60 days following pollination, the growth in pod dry weight is gradual and peaks after around 100 days before it declines. To put it another way, **under-fertilization could result in less seed production**.

Consequently, fertilization serves as a gauge for the number of seeds within a pod.

According to the weather and rainfall patterns in a given area, harvest times can vary greatly.[6]

Temperatures higher than 23 °C restrict flowering in places with no clearly defined dry season, leading to low yields seven months after the start of the season.

Rainfall distribution has a six to seven-month delay between the beginning of the rains and the ripening of the fruit in warmer regions.

Different clones have different responses to changes in air temperature when it comes to fruit growth. As a result, fruit ripening time, cherelle wilt losses, final pod size, bean size, and lipid content are affected.[7]

Crop yield

There is an estimated 56 t ha-1 annual biomass (above ground) productivity for cocoa (Corley, 1983).

The maximum seed yield is expected to be 11 t ha1 based on an assumed harvest index of 0.20.

Compared to the greatest recorded yield of 4.4t ha-1 (without shade) and the best commercial yields of 1.5–2.5t ha-1, this is a significant increase in yield.[8]

Additionally, excessive shading from trees limits photosynthesis and competition for water. Higher leaf temperatures, particularly when trees grow without shade, with low air temperatures make foliage vulnerable to wind damage.

It has been shown that selective breeding for efficient biomass allocation to the yield component can increase cocoa yields.[9]

In an 18-month field experiment, the researchers discovered some variation in dry bean yields among the 12 genotypes examined, ranging from 200 to 1400 kg ha-[1].

From 11.1 cm^2 to 27.6 cm^2, the trunk's cross-sectional area grew at a rate of between 0.08 kg/ cm^2 during the same period.

Moreover, one-third of the pod biomass in seven of the clones studied contained beans.

Chapter 3

SELF-INCOMPATIBILITY IN CACAO

Pollination in cacao is complicated because most of the trees cannot pollinate themselves. Therefore, most cacaos undergo cross-pollination.

The pollinators have to travel around and pollinate with pollen from different trees. Unlike other tree crops pollinated by bees, natural pollination in cacao is just a chance or possibility

Theobroma cacao has a genetic system for self-incompatibility that is unlike any other in flowering plants. A single S allele is responsible for the sporophytic system in both male and

female organs, according to the findings of Knight and Rogers.[1]

Although pollen-tube research indicated that incompatible crosses did not hinder pollen-tube growth, the ovules could liberate the pollen-tube contents, which is why they failed to set fruit.

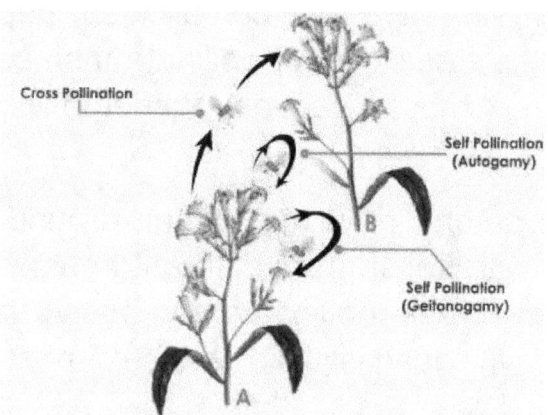

The problem: Pollination rates in cacao tend to be 25 to 37.5%. This is the case of well-managed cacao farms.

Nonetheless, some traditional cacao farms have pollination rates of just 3 to 5% percent. This explains why most traditional cacao farms have a very low yield.

Recent research suggests that self-pollen identification may emerge sooner than originally thought. Some researchers discovered that until the concentration of CO_2 was raised, pollen grains of the self-incompatible clone IMC 30 would not germinate following self-pollination and that self-incompatibility could be overcome by applying CO_2 to self-pollinated cacao flowers.[2]

There was no fusion of the gametes in untreated, self-pollinated flowers. Self-incompatibility in cacao may be caused by two mechanisms: one occurring during pollen germination (pollen-stigma contact) and the other occurring after gametic fusion.[2]

Isocitrate dehydrogenase (IDH), malate dehydrogenase (MDH), and acid phosphatase (AP) activity were found to be good predictors of self-incompatible Theobroma clones.[3]

Chapter 4

THE CACAO POLLINATORS

➡   The little *Forcipomyia midge* is a particular genus of fly that pollinates the cacao flowers.

➡   It's 1-3mm tiny. However, only flies between 2-3mm can pollinate the cacao flowers because the others are too small to reach the important bits.

➡   This implies that bees and other insects bigger than 3mm cannot pollinate cacao flowers.

Source: https://www.nationalgeographic.com/animals/article/animals-food-thanksgiving-environment

⬛ The *Forcipomyia midge (the pollinators)* live in the decaying matter at the base of the cacao trees and swarm in huge clouds all around the lower branches.

Source: https://cacaopollination.com/cacao-pollinators/

Causes of low yield in cacao include scarcity of the main pollinators (*Ceratopogonid midges*), especially in the dry season

Why Pollination in Cacao is Hard

Apart from the earlier discussed factors which make pollination difficult, the characteristics of the cocoa flower seem to make it unattractive to many potential pollinators. Therefore only insects that evolved with the cacao will successfully pollinate it.

The midges contact the pollen-bearing stamens of the flowers and then transfer this pollen to the pistil of another flower, perhaps some meters away.

Because the cacao flowers are resistant to self-pollination, just like most of the world's flowering plants, most of the pollen distributed by midges comes, obviously, from the flowers of neighboring trees; however, unlike bees or butterflies, they don't tend to roam far.

The ovary of each cacao flower can recognize its own pollen, and the tree rejects that flower, causing it to wither and drop from the tree.

Only pollen from a different tree will be allowed to form a cacao pod.

This can lead to very low amounts of fruit set. Recent research into flowering and pollination has led to some very interesting discoveries that will be impacting growers' decisions in the coming decades.

It turns out that most cacao trees will allow at least a few flowers to self-pollinate, while some varieties permit as much as 50 percent of them to self-pollinate.[1]

Environmental factors and tree stress can change these percentages, too. As long as the propagation of cacao trees occurs via softwood cuttings and not by seed, genetic flaws will not be allowed to develop.

Research findings

Optimum floral production occurs at temperature 22.5 °C, light intensity 91.8 Fc, and rainfall of 141.1 mm per month; however, rainfall could be the most critical factor in the floral phenology.

Flower stability is affected by seasons and pollination. The midges visit the farm more than other insects. Only midges could carry 60.1% of pollen needed for pollination.[2]

Chapter 5

WEEDING & PRUNING FOR EFFECTIVE POLLINATION

should be done at the beginning of the rainy season. Weeding can be done 1-3 times a year, depending on the age of the cacao

The use of weedicides is highly discouraged as they have harmful effects on pollinators and soil microbes.

Manual weeding is highly recommended to sustain the growth of pollinators.

Pruning the Cacao Farm

Pruning should be done after weeding. It reduces the incidence of diseases and pests; ensures proper ventilation and light distribution for proper photosynthesis; It also opens up branches to flowering.

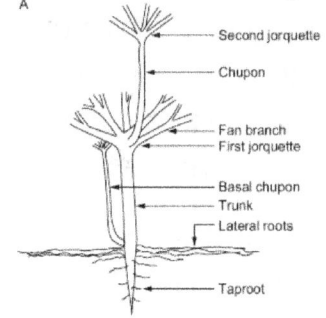

A
Second jorquette
Chupon
Fan branch
First jorquette
Basal chupon
Trunk
Lateral roots
Taproot

Pruning tools

The basic aim of pruning cocoa trees is to stimulate flowering and facilitate harvesting.

Young plants should be allowed to develop a jorquette at the height of at least **1.5 meters.** However, the jorquette height varies significantly from tree to tree.

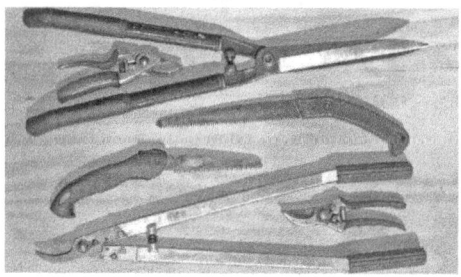

When trees are trained to form branches at the height of 1.5m above ground level, flowering and fruit formation can be increased by about 25% compared to untrained trees with branches at the ground level.

This is due to proper farm ventilation, breaking the microclimate for disease build-up, triggering flower buds, and more surface area per tree for flower formation.

It has been found that increasing light intensity decreases the jorquette-height. If a jorquette is considered too low, it can be cut off.

The strongest of the re-growing chupon can be selected, and all others removed.

In due course, this chupon will produce a jorquette at a higher level. Vegetatively propagated plants generally form a jorquette at a lower level.

25

How to Prune

Phase 1: Pruning to Lower Height

Phase 2: Pruning to Open Crown

Phase 3: Pruning to Open Rows

  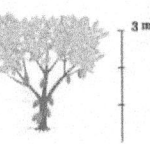

Source: https://www.technoserve.org/blog/revolutionizing-cocoa-farming-in-peru/

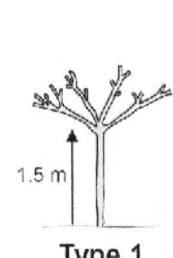

Type 1

Cocoa tree managed following usual technical recommendations

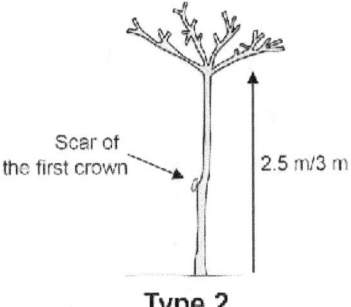

Type 2

Type 1 cocoa tree iterated one or more times

Source: https://www.fao.org/3/ad220e/ad220e04.htm

Proper pruning

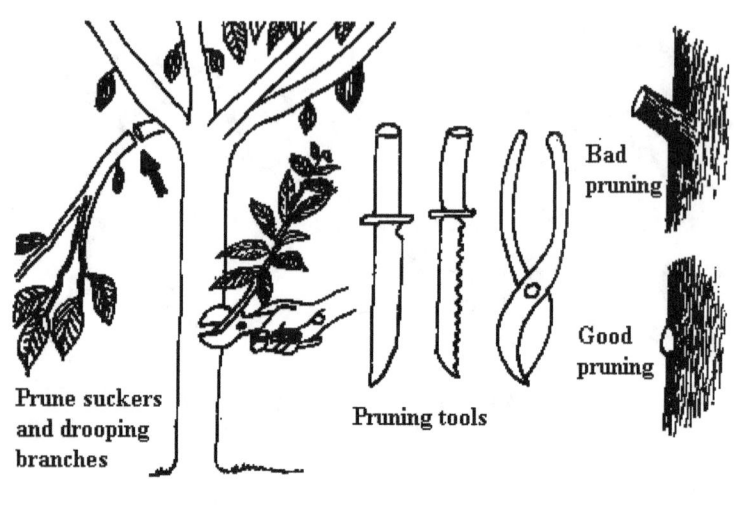

Prune suckers and drooping branches

Pruning tools

Bad pruning

Good pruning

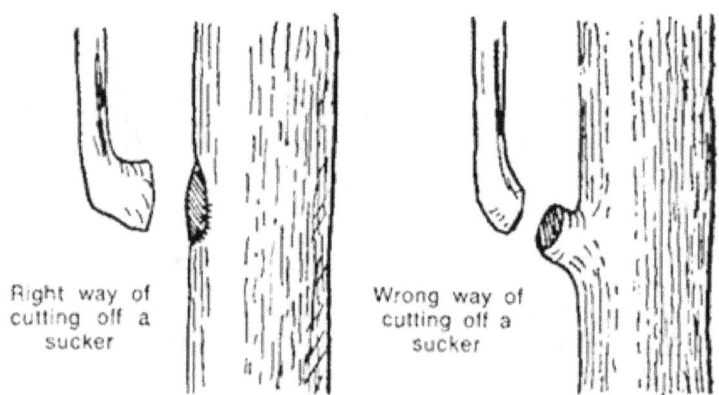

Right way of cutting off a sucker

Wrong way of cutting off a sucker

Source: https://www.fao.org/3/ad220e/ad220e04.htm

Chapter 6

SHADE MANAGEMENT FOR EFFECTIVE POLLINATION

A 70% shade is best suited for the early growth and development of the young cocoa plant.

This shade should be gradually reduced to 50% for 3-year-old cacao trees and 25% for trees over 5 to 7 years old.

Two types of shade trees are used in cocoa: Temporary shade and, Permanent shade

If shade is not properly managed, flowering, pollination, and fruit formation will be reduced.

Diseases and pests will instead proliferate on your farm.

Shade trees have limited benefits for soil fertility in cocoa agroforests.

Inferring from this, shading does not improve soil fertility in cacao plantations, and therefore needs to be well managed.[1]

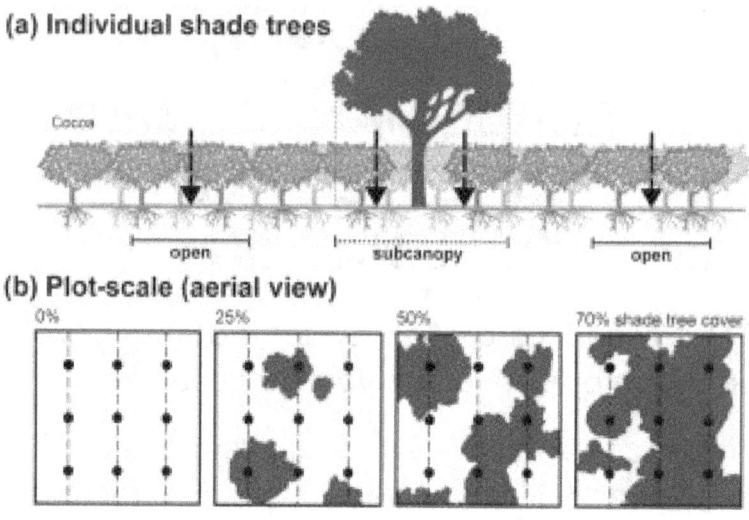

(a) Individual shade trees

Cocoa

open | subcanopy | open

(b) Plot-scale (aerial view)

0%   25%   50%   70% shade tree cover

Source: https://www.sciencedirect.com/science/article/abs/pii/S0167880917301615

Cocoa growth is lower under individual shade trees and decreases with increasing shade-tree cover in plots, and cocoa yields also decrease with increasing shade-tree cover.

If cacao is intercropped with other trees crops, you ensure that the cacao trees are not completely submerged in the shade. There needs to be some sunlight penetration through the leaves to the bed of the farm.

Two Types of Shading

Temporary shade can be provided by food crops such as bananas or plantain. You can maintain temporary shade for a period of about one to three years.

Permanent shade forms a canopy over the mature cocoa plants. Fruit trees, plantation trees, and timber trees can be used for permanent shading. Both permanent and temporal shading trees/ crops should be planted at least one year before the young cocoa trees are planted. In this way, the cacao will have shading upon planting.

Effects of shade-tree species and spacing on soil and leaf nutrient concentrations in cocoa plantations at 8 years after establishment

Cacao can be planted with a non-legume tree-like *Gliricidia sepium*. The shade-tree spacing 12m × 12m is recommended. The growth of shade trees needs to be managed so they don't become a nuisance to flowering and fruiting.

Soil total carbon (TC) and total nitrogen (TN) will significantly increase in Theobroma + Gliricidia plantation.

The Theobroma + Gliricidia plantation also increases soil water extractable phosphorus (P) when regularly pruning.[2]

Chapter 7

NUTRITION MANAGEMENT FOR EFFECTIVE FLOWERING & POLLINATION

Fertilizer Application

Choice of fertilizer when it comes to proper so management is all about obeying the **4R strategy**:

The **R**ight type, **R**ight amount, **R**ight timing, and **R**ight placement.[1]

Heavily shaded fields do not respond as well to fertilizer as fields with minimum shade.

Use fertilizers as recommended based on the results of a soil test. Use Nitrogen, Phosphorus, and Potassium (NPK) fertilizers if the soil is lacking in these nutrients.

The amount and type of nutrients required vary with the age of the plant.

RECOMMENDATIONS

Newly Transplanted Cocoa:

Newly transplanted young cacao plants require **more phosphorus** for good root development.

If the soil test shows that the soil does not have enough phosphorus, use **NPK** fertilizer high in phosphorus, for example, **12:24:12** at the rate of 0.1 kg per plant. Fertilizer application can be done 3-4 times per year.[2]

Young Trees (1-3 years)

At this age, the young cacao trees require more nitrogen for vigorous growth. If the soil lacks nitrogen, young trees should receive two applications of a fertilizer high in nitrogen for the development of shoots.

For example, apply **30:10:10 NPK** fertilizer at the rate of 0.2 kg per tree, twice per year for the first three years.

Actively Growing Trees (over 3 years)

If the soil lacks nitrogen, apply a high percentage of nitrogen fertilizer such as **30:10:10 NPK** fertilizer at the rate of **0.2 kg to 0.4 kg** per tree twice per year.

Flowering and Fruiting Trees

Flowering and fruiting trees have a high requirement for **Nitrogen and Potassium**. If these are lacking in the soil, a fertilizer such as

16:8:24 NPK is recommended at the rate of 0.4 kg to 1.2 kg twice per year.

FERTILIZER PLACEMENT

Fertilizers should be placed in circular bands around the trees for young cocoa plants, 5.1 cm - 10.2 cm (2 to 2 inches) away from the plant.

You could also use the canopy extent to draw a circle around the plant for fertilizer placement. Dig a circular trench around the plant, place the fertilizer and cover it up.

As the trees mature, the fertilizer should be applied in circular bands located further away from the trunk at the drip circle.

In mature trees, it is best to broadcast fertilizer on the soil surface throughout the field.

One application can be made at the <u>beginning of the rainy season</u> and the other three to four months later after rainfall or when soil is not dry

Chapter 8

ENHANCING NATURAL POLLINATION IN CACAO

Cross-section of the Cacao Flower

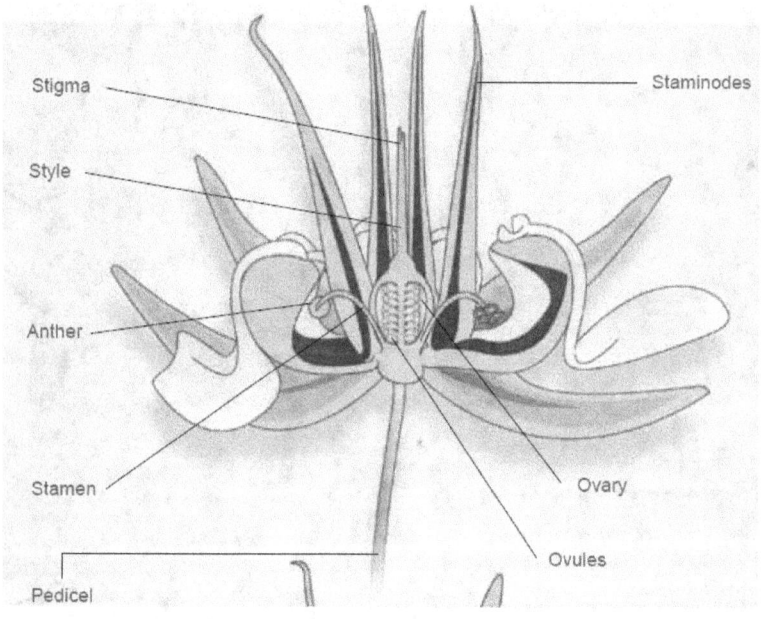

Stigma — Staminodes

Style

Anther

Stamen — Ovary

Ovules

Pedicel

Cacao Inflorescence

The cacao flowers immerge in groups directly from the main stem or older branches.

The flowers appear at locations on the stem and branches, which were originally leaf axils. Each flower has five unique pink sepals, five smaller yellowish pouch-looking petals, an outer whorl of five staminodes, and an inner whorl of five double stamens. Each stamen bears up to four anthers.

The staminodes are about two times taller than the style and form a "fence" around the style.

The stamens are curled, so the anthers usually develop inside the petal pouches. The ovary consists of five united carpels. Each carpel has four to 12 locules and one style that has several linear stigmatic lobes

The flower produces no nectar and has no discernible scent. However, further research has revealed two types of microscopic nectars: (1) the cylindrical multicellular ones, 60 to 450 microns in size, on the pedicels, sepals, and ovaries, and

(2) the conical unicellar ones are 20 to 25 microns in size, located on the "guidelines" of the petals and the staminodia.

They secrete nectar, tend to attract male mosquitoes and lepidopterous insects.

The cacao flower opens about dawn, and the anthers open just before sunrise. The stigma is usually pollinated within 2 to 3 hours of the morning. However, they are receptive from sunrise to sunset.[1]

The stigma is receptive to pollen along its whole length, and not only at the apex as in most flowers.

If the flower is not pollinated, it usually sheds or falls off within a day. Effective pollination happens before 10 am.

Natural Pollinators

Many works of literature credit up to 60% of cacao pollination to midges, especially *Forcipomyia quasiingrami* Macfie and *Lasiohela nana* Macfie Other literature shares that ants (*Crematogaster spp.*), aphids (*Aphis gossypii* Glover and *Toxoptera spp.*), thrips (*Frankliniella parvula* Hood), and unidentified tiny wild bees also contribute to pollination on cacao.

It is reported, after a study of albino trees, that a considerable proportion of pollination takes place across two intervening trees, though less than over shorter distances. This would indicate an agent with considerable movement between trees.[2]

In a study on ants and aphids, the researcher found that of 5 percent of the flowers on trees not infested by ants and aphids, only 0.3 percent set fruit; whereas, on trees heavily infested by these insects, 35 percent of the flowers were pollinated and 2 percent set. At the same time, 5 percent of the hand-pollinated flowers set fruit.

Apparently, wherever cacao is grown, the lack of adequate pollination is a strongly limiting factor in the production of the beans.

It has been documented that most pollination occurs 2 to 3 hours after dawn with a second much smaller peak in the afternoon, but only 4to 9 percent of the flowers ever get pollinated. Fruit setting may not happen if pollinated too late or with incompatible pollen.[3]

Because some cacao trees are self-incompatible - some are sterile.[4] Due to sterility, many of the flowers would be shed.

Knoke and Saunders tried a mist blower for the mechanical transfer of pollen but achieved uneconomical success. This means wind does not really contribute significantly to pollination.[5]

A minimum of 60 pollen grains is necessary to set the highest number of seeds per pod.

The quality of pollination can depend on two factors:

1. The degree of pollen compatibility and

2. The number of pollen grains deposited on the stigma.[6]

Fertilization to seed ratio

For example, if a fruit has 40 seeds, it means that 40 pollen grains fertilized 40 ovules. This means seed formation is dependent, and pollination and fertilization

Enhancing Natural Pollination

Recommendations

1. Do not remove the leaves that fall to the ground from cacao and shade trees. Leave a layer of leaves on the ground to provide a home for the pollinators.

2. Insecticides or other toxic substances kill the pollinating midges and other insects that

contribute to pollination. As much as possible, reduce using chemicals in the cacao plantation.

The use of pesticides should be used before flowering. Precautions should be taken not to spray the dry decay leaves on the soil.

3. Grow shade trees, fruit trees, and bananas throughout the cacao grove so that temperatures remain cool and the midges can reproduce, grow, fly and pollinate the cacao. Do not completely eliminate shade trees and retain rainwater because this is another place where the pollinating midges lay their eggs.

4. Provide more places for the midges to lay their eggs, cut the stems of banana or plantain trees into disk-shaped pieces or slices around 5 centimeters thick,

Distribute these in different parts of the cacao plantation, near the trunks of the cacao trees.

It is best to keep those banana or plantain stem slices throughout the year, especially during the months when the cacao trees are flowering.

Chapter 9

THE LIFE CYCLE OF MIGDES & HOW IT IMPACTS CACAO POLLINATION

Adults: Biting midges grow from egg to larva to pupa, and finally to the adult stage. The complete cycle can occur in **two to six weeks**. The life cycle also depends on the species and environmental conditions.

The adults are most abundant near productive breeding sites but will disperse to mate and to feed. The mean distance for female flight is 2 km, less than half of that distance for males.

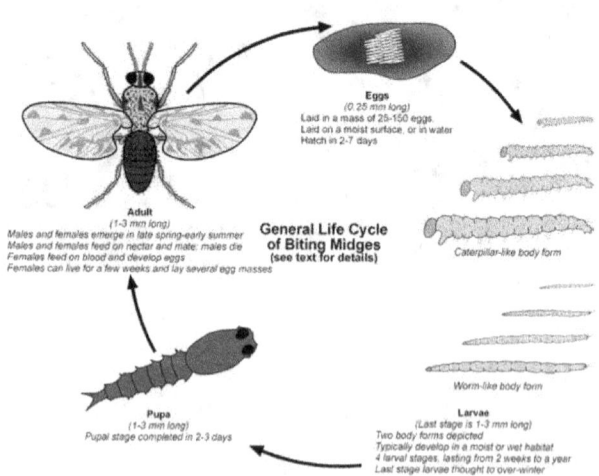

Source: https://extension.entm.purdue.edu/publichealth/insects/bitingmidge.html

The female lays eggs on various moist surfaces and hatch in 2-7 days. There are four larval stages (Figure 1), with larval development completed in about **2 weeks** to a year or more, depending on temperature and food supply.[1]

The pupal stage typically is formed in the same site as the last larval stage, and adults emerge in **2-3 days**.

Adults can **live for 2 to 7 weeks**. Laboratory and field studies suggest that biting midges may complete two or more generations per calendar year.

Biting midges undergo complete metamorphosis. This means the last larval stage molts into a non-feeding pupal stage that eventually transforms into a winged adult. Relatively few species have been studied, and the account below is based largely on pest species that have been reared in captivity.

Male and female biting midges feed on plant sap and nectar, the primary energy sources for flight and increased females' longevity. Egg production requires a protein source, obtained either from the body fluids of small insects or vertebrate blood. Male biting midges are not attracted to vertebrates, and their mouthparts are not capable of biting.

Females of pest species feed primarily on mammals, but birds, reptiles, and amphibians also are a source of blood meals. Some species are host-specific, but others are opportunistic, feeding on a variety of vertebrates that they encounter, usually in response to carbon dioxide emitted by the host.

Different species of biting midges have their peak feeding periods at different times of the day. For example, females of *Leptoconops species* feed during daylight.

In contrast, females of *Culicoides species* typically do not feed until dusk and continue feeding at night.[1]

Implications of their life cycle on pollination

The life cycle of midges shows that midges live up to 7 weeks. Within this period, they carry out pollination in about 4 weeks, then die. A new generation takes over the pollination task.

This implies that the use of pesticides on the cacao plantation, if ever, should be in about 2 months to flowering. In this way, if large colonies of adult midges are killed, the eggs in debris, leaves, husks, banana pseudostems will hatch and grow to pollinate the farm. Extra precautions and calculations need to be done so as not to risk pollination of the cacao farm.

During pod formation, the use of insecticide or fungicide could be selective on areas of the farm where it's needed. Sections with flowering flowers should not be sprayed.

In this recent study, fresh cocoa pods were left to decay in the cacao farm. It was observed that the adult midge preferred to feed and lived in the substrate through the reproductive cycle.

The total midge population in the cocoa pod was 3 to 4 times higher than the banana pseudostems.

The data suggest that increasing the breeding sites with the augmentation of cocoa pod substrates can significantly increase the midge population dynamics in the cacao farm and enhance new pods development.[2]

Table 1. Population of midges harvested on cacao farm in Gran Couva

Substrate type	Average male	Average female	Average midges	Total midges
Cacao pods	123.9	192.2	316.1	5660
Banana pseudostem	37.61	71.8	109.5	1885
Cacao leaf litter	1.1	2.81	3.7	65
$\bar{x} \pm SE$	54 ± 36.4	88 ± 55.3	143 ± 91.7	2537 ± 1648

Source: Bridgemohan et al. (2017)

It could be seen from Table 2 that during the wet months of July/August, the number of midges caught in the suction traps located in the areas of the banana pseudostem, and cocoa pod increased, compared to the litter substrate. Similarly, the cocoa leaf litter was not significantly different from pods or pseudostems in August.

The table portrays that if farmers can keep their farms moist with stagnant waters, the activity and proliferation of pollinating midge will be enhanced.

It could also be seen that the contribution of leaf litter, decay cacao pods, and rotten banana stems are equally important for the growth of the midges.

Table 2. Population of midges harvested on cacao farm in Centino in different months

Months	Cacao leaf litter	Cocoa pods	Banana pseudostems	x̄ [SE]
March	4.75	4.25	4.75	4.6±[0.17]
April	5	4.25	2.75	4.0±[0.66]
May	3.5	3.25	1.5	2.8±[0.63]
June	2.75	1.75	2	2.2±[0.30]
July	5	8.25	7	6.8±[0.95]
August	11.5	9.5	11.75	10.9±[0.71]
x̄ ± SE	5.4 ± 1.27	5.2 ± 1.23	5.0 ± 1.59	

Sources: Bridgemohan et al. (2017)

Chapter 10

ARTIFICIAL OR HAND POLLINATION

Hand pollination can be done like this: pick a flower from the father tree, remove the petals, revealing the anthers, and then rub it on the stigma or style of the flower of the mother tree.

How to Hand Pollinate Cacao

Step 1

Select the mother flower you will pollinate.

Step 2

Select the father flower that will pollinate the mother flower.

Step 3

With tweezers remove the petals from the father flowers to uncover the anthers where the pollen is.

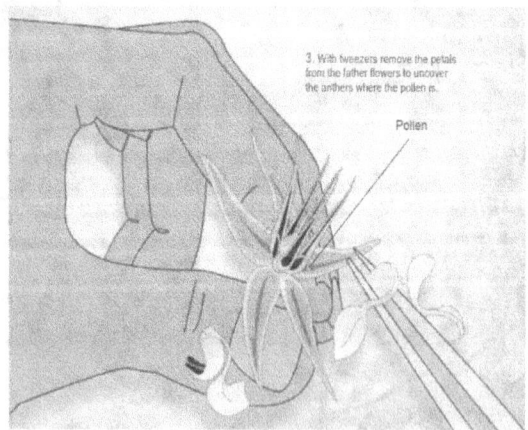

Source: https://www.worldcocoafoundation.org/wp-

Step 4

Hold the mother Staminodes flower carefully, and with the tweezers, remove the staminodes to uncover the stigma.

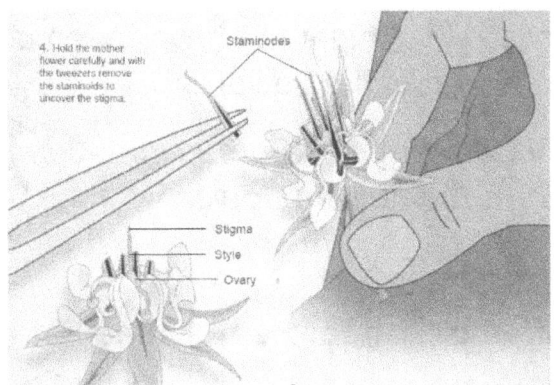

Source: https://www.worldcocoafoundation.org/wp-

Step 5

Rub the anthers of the father flower on the stigma of the mother flower so that the pollen will stick to it.

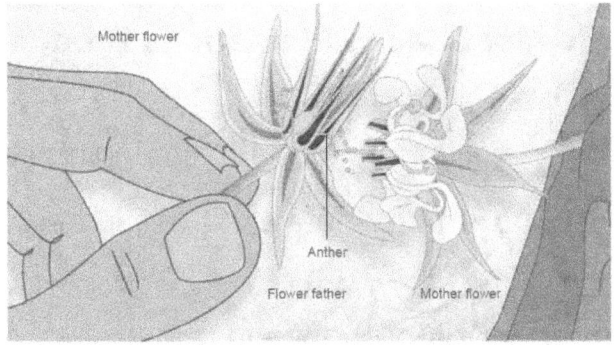

Source: https://www.worldcocoafoundation.org/wp-

Pollination method - Option 2

Use the forceps to remove the anther (male) with the pollen grains and rob it on the stigma(female) of the mother tree.

In this case, you might not have to remove any other floral part of the flower

It is recommended that you rub 2-3 male flowers on the female flower.

Recommendation for Established Farms

1. Form farmers' groups/ teams to pollinate one farm at a time.

2. You could also hire labor, train them to hand pollinate the farm.

The ROI is usually over 5 times higher than the cost of labor in pollination

Why Do Most Cacao Plantations Produce So Little?

1. Too few pollinators in the cacao plantation.

2. A lot of **variability** between trees in the cacao plantation.

3. **Incompatibility** between the trees in the cacao plantation.

4. Lack of maintenance in the cacao plantation.

5. Lots of pests and diseases.

Chapter 11

VARIABILITY & INCOMPATIBILITY BETWEEN CACAO TREES

Variability

Variability in the cacao plantation is a direct result of sexual reproduction in cacao.

The problem is that when variability is not controlled, the majority of the trees produce only a few fruits.

When the variability is not controlled, most of the harvest is produced by only a few trees: 70 out of every 100 kilos are produced by 30 of every 100 trees that grow in the cacao plantation.

In order words, 70 percent of the harvest is produced by 30 percent of the trees in the plantation.

If you manage to turn all your trees into good producers – after the harvest, you could take off on a two-week vacation with your family and

stay at a luxury hotel.

The good news is….

You can increase the number of trees that are good producers in the cacao plantation. There are techniques for doing that.

Compatible

Cacao trees are considered compatible when two trees are able to pollinate well together.

When the pollen of one tree can fertilize the flowers of another cacao tree, we say that the two trees are compatible with each other.

But if the fruits and seeds routinely fail to develop from a particular cross, we say that these trees are incompatible with each other.

Self-incompatibility

Some cacao trees can fertilize themselves, while others can't. When the pollen of a tree can fertilize the flowers of the same tree, we say that the tree is **self-compatible.** If it cannot, then we say that the tree is **self-incompatible.**

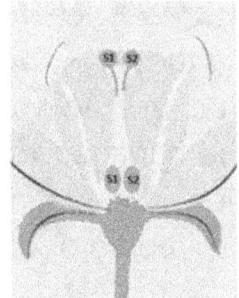 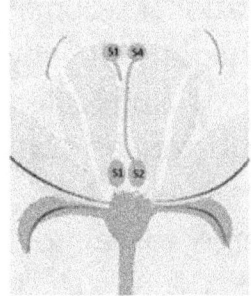 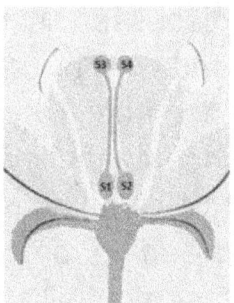

https://www.quora.com/Why-do-some-flower-bearing-plants-have-no-fruit

Caution

You must pay a lot of attention to the issue of compatibility and incompatibility because it can really affect the production of a cacao plantation.

For example, some trees produce flowers but do not produce fruit. Often this is caused by incompatibility.

If the trees in a cacao plantation are not compatible with each other or themselves, there will be less fertilization.

The plantation would produce many flowers but few pods.

The incompatibility system in Amazon cacao is discussed in relation to dominant gene action and the "antigen-antibody" theory of incompatibility. It is suggested that the active incompatibility substances could be confined to the cytoplasm of ovules and pollen in genetic breeding programs to reduce the problem of incompatibility

HOW TO MANAGE INCOMPATIBILITY IN CACAO TREES

Practical Farm Management Suggestions

- 1. Divide your farm into equal blocks

- 2. Closely identify areas that are high yielding and areas that are not

- 3. Select flowers from the areas that are high yielding to hand pollinate areas that are low yielding

Most of the time, low pod formation could be traced to incompatibility problems

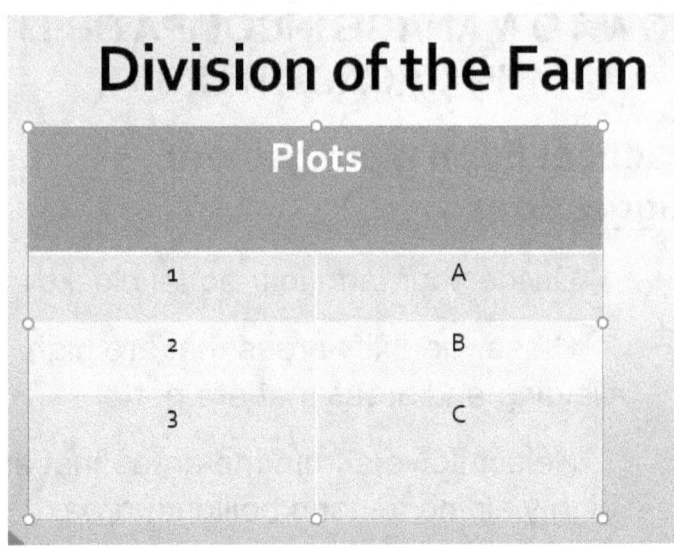

Division of the Farm

Plots	
1	A
2	B
3	C

Pollen from Block A could be used to pollinate trees in Block 2. In the same way, pollen from Block C could be used to pollinate trees in Block 1. Hand pollinating your farm will always pay off, so you may want to mobilize a team to carry it out.

Remember ....

Your farm is your business. You don't wait for results, but you create the results you want on your farm

The solution to incompatibility is to introduce trees that are compatible with the other trees on the plantation and with themselves. Aside, enhance cross-pollination more than self-pollination.

In other words, they should be self-compatible as well as compatible with nearby trees.

If there are self-incompatible trees and incompatible with other trees, you may want to cut them down and replace them totally.

When buying seeds….

You must think about compatibility when you're buying seed.

For many farmers, when they want to establish new cacao plantations, don't pay enough attention to where they obtain the seed.

When choosing seeds…

Usually, when you want seeds, you ask your neighbor for seeds from a very productive tree that produces many pods, with large seeds and other good traits.

But, it's also possible that those seeds could produce trees that are not compatible with other trees in the cacao plantation.

Get certified seeds…

In that case, it's advisable to find out where to obtain certified seed.

In those places, they select trees with good traits and without problems of incompatibility.

Remember…..

…that even if you cross selected trees that are compatible, the offspring of these trees will still not be exactly like their parents.

They will also be different from each other.

How To Resolve Variability and Incompatibility

The best way to resolve the problems of variability and incompatibility is by using **reproduction techniques** that do not rely on the compatibility of the sex cells

Asexual Reproduction Techniques in Cacao

Several asexual reproduction techniques are used in cacao, such as grafting, layering, rooting cuttings, and somatic embryogenesis.

https://www.foto-grafo.de/Cocoa/cocoa_4.html

Chapter 12

CACAO FRUIT ABORTION: CAUSES AND PREVENTION

It is estimated that only 0.5–5% of cacao blossoms are pollinated naturally, despite their abundance. Cherelles grow from pollinated flowers. Despite many flowers and pod sets, only a small number of cherelles mature into pods.

Cherelle wilt can cause the death of up to 75% of a crop. Insect, and fungal pests can all destroy Cherelles.[1]

Growth of Cacao Pods and Flowers Cacao

The most developed stems and branches have the heaviest flowering and pod production. [2.] Within 32 hours of anthesis (flower opening), the unpollinated flowers normally drop off.[3]

Hormones and flower drop

Hormones are involved in abscission (flower drop), which is necessary for aging because ABA levels rise dramatically just before it occurs.

Ethylene also plays a supportive function in this process.[3] The first symptoms of Cherelle development appear 6–8 days after successful Pollination.[4]

Cherelle wilt begins with a loss of water balance in the pod, resulting in a drop in pod turgor pressure.[5] Cherelle wilt causes a mucilage-like substance in the cherelle peduncle to clog xylem vessels, which is the first apparent alteration.[6]

These limit the growth and development of the pods by limiting the flow of water into them.

Pods appear to lose turgor pressure in the early stages of cherelle wilt, and the cherelle yellows over a week after that.[6] Sometimes, the yellowing starts at the pistil end of the pod and then moves throughout it. The pods turn a dark brownish black after turning yellow.

This results in mummification of the cherelle since the size of existing vascular bundles increases, but new bundles aren't formed as quickly, as well as lignification of the middle pericarp.[7] The tree's cherelles are mummified and still attached to it. Cherelle wilt occurs in two stages or flushes referred to as first wilt and second wilt.[8]

This stage begins about 50 days after pollination and concludes when the seed endosperm cell walls begin to form.

A cacao tree's yield is thought to be correlated with the number of flowers pollinated successfully.

Compared to naturally pollinated trees, hand-pollinated trees produced more pods. Still, these trees also had a higher degree of cherelle wilt, which meant that the number of ripe pods in hand-pollinated and normally pollinated varieties was the same.[9]

Cherelle wilt kills pods by interfering with the pollen's ability to stick to the plant. *T. cacao* flowers have a two-stage incompatibility reaction that is regulated by phytohormones. Pollination and gamete fusion are key factors in fruit set.[9] Pollination causes an early release of abscisic acid (ABA) during the pollen–stigma contact, according to Baker et al.

The cacao flower's hormonal composition may have a function in regulating incompatibility with oneself.[11] This ABA release happens during the reaction of self-incompatibility and is likely essential for later abscission of floral parts.

Unpollinated flowers have no ABA release. In unpollinated flowers, the rising amounts of ethylene are not countered by rising auxin levels, resulting in the characteristic abscission[12]. After the pollen tube comes into touch with the ovule, the second stage begins. The auxins and ethylene levels change as a result of this touch, and this is what determines whether the flower turns into a fruit or falls off the tree[12].

When the levels of ethylene and ABA are high in flowers, which is frequent in self-incompatible flowers, abscission is most likely to occur. Abscission is overcome when the rising quantities of auxin balance the rising levels of ethylene, allowing fruit set to occur.[12] Other researchers have observed fruit set on self-incompatible trees;[13] nevertheless, the tree frequently aborts these pods.[14]

Studies suggest certain self-incompatible clones have a low open pollination pod set but a higher mature pod seed set.[15]

Potential Causes of Cherelle Wilt

Leaf flushing

Studies have shown that during or shortly after bouts of leaf flushing, there is an increase in cherelle wilt.[16,17] According to one theory, because the tree cannot produce enough energy to support enhanced leaf growth and pod development, it loses part of its young pods.

Root shoot balance

There is a greater risk of cherelle disease developing on branches as opposed to the main stem of the tree when cherelles are located there.[18] A connection exists between plant sugar and water transport, as evidenced by the occurrence of wilt. Pods on smaller branches further out from the roots were found to be more susceptible to wilting, according to a study conducted in Trinidad.[17]

Age of cacao trees

Cholera wilt is more common in young cocoa plants, but it becomes less common as the trees grow older.

Despite the increased number of leaves and improved photosynthetic capability, this is because young trees lack sufficient food reserves to promote pod development [17]

Phytohormonal deficiency

As with apple wilt, hormone deficit may be to blame for cherelle wilt in cacao. As previously mentioned, a lack of hormone production in the endosperm may result in decreased water and sugar transport in pods, leading to the wilt condition.[8]

Ethephon was found to cause wilt in early fruit when applied to the pedicels, whereas gibberellic acid treatment prevented wilt in cherelles 10–50 days old.[19]

More flowers and fruit sets were observed on the trees after using the gibberellic acid inhibitor chlorocholine chloride.[20] Application of benzyladenine with acetic acid or gibberellic acid to Amazonian cacao trees studied in India reduced the symptoms of cherelle wilt.[18]

Plant growth regulator application has less effect on cherelle leaf spot than treatment, environmental factors, and tree clone.

Effects of Environment on Cherelle Wilt

Cherelle intensity

Cherrelle wilt is strongly influenced by the farm environment in which it is grown. Cherelle wilt will occur if it rains too much or if it is too hot. Cherelle wilt is less common in areas of the flower cushions with fewer flowers. [21]

For the tree to manage fruit production costs, it must be able to control both the pod and flower production costs.

Shading

Shade is one of the environmental elements that affect the development of pods and the occurrence of cherelle wilt.

There's a growing debate about whether cacao grown in the sun or the shade produces better results. The effect of shade on cherelle wilt may become relevant for output as cacao is planted in larger, shade-free farms. Reduced humidity in the under-canopy can inhibit the spread and sporulation of stramenopile and fungal infections, which is why shade management is so important for treating cacao diseases. Numerous studies have been done on the effects of shade on the Cherelle wilt tree that show that soil fertility and the subsequent nutritional level of the tree are more important than the actual shadow.

Asomaning and colleagues conducted experiments in Ghana to see how shade affects flowering and fruiting, and found that fertilised trees in low shade produced more pods than unfertilized trees in high shade or no shade, supporting the theory that fertilized trees are more resistant to wilt in any environment.[22] Cherelle wilt was not impacted by shade in diverse agroforestry systems in Central Sulawesi, Indonesia, where researchers worked with hand-pollinated pods. Because nitrogen-fixing legumes were more abundant under forest trees than under shade trees, the nutritional levels of the tree had an impact on cherelle wilt.[14]

As previously mentioned, reducing shade is likely what caused the accompanying increase in cocoa production (Wood and Lass 2001). Additionally, as previously indicated, an increase in photosynthates helps assist pod development by supplying more nutrients. In order to compete for soil minerals, cacao plants grown under significant shading must also compete with shadow trees for nutrient availability.[23]

Rather than being a shade response, cherelle wilt is more of a response to soil health and nutritional levels than a response to shade.

Detection & inspection

Look for the tree's bare branches that are standing straight up, as well as the buds that have peculiar branching. Cherelles should be black with a white fungal growth on them if you want to find them. This fungus is rarely the cause of the cherelles' demise; instead, it appears after the flowers have begun to wilt. (*Phyophthora palmivora*, the organism that causes black pod disease, is not the same thing.) Instead of browning or turning black, the water mould generates white growth on green (or red) cherry tomatoes and peas, as in this case.

Management

Cultural Control

Cocoa tree health can be improved to limit the spread of cherelle wilt. Fertilizers and mulch are the two most critical tools for accomplishing this. Cherelles are less likely to wilt when nutrients are abundant.

Simple rules to follow.

Sunlight is required for the production of plant meals, but it must also be balanced with sunscald to prevent foliage damage.

The number of trees in an area

Avoid placing trees too close together when you're planting them. Tree crowding increases the risk of Cherelle wilt due to competition for nutrients, water, and light.

Moisture in the soil. The risk of cherelle wilt is higher in dry or wet soils, so use drains or mulches to keep them moist.

Fungi and insects can be controlled by using pesticides. Unless there is an epidemic of black pod sickness, this is likely of minor importance.

Chapter 13

ENHANCING DOUBLE MAIN HARVEST IN CACAO

All things being equal, a common denominator in cacao yield is the presence of moisture and nutrients in the soil. These are two factors that separate the main season/crop from the minor season/crop.

To enhance these two main factors, farmers will need to raise soil moisture and nutrient availability to the cacao trees throughout the year. Ensuring this will as well raise the survival rate of the cacao trees.

Cacao Responses to irrigation

Surprisingly, few irrigation studies have been conducted considering cocoa's apparent susceptibility to drought, but this is understandable given how unlikely commercial-scale irrigation is.

A lack of moisture in the atmosphere during the dry season was cited as a potential limiting issue to cacao yield.

In a three-year irrigation experiment in Ghana, mature Amelonado trees yielded an increase of 12, 17, and 40% in cocoa production by keeping the soil close to field capacity. Due to a lack of 'dry air' limitations, these increases were less than expected at the time of their implementation. There were no yields stated in absolute terms.[1]

A second irrigation experiment in Malawi was carried out.[2] More than 1,650 mm of water was poured throughout a protracted dry season in two years with an average annual rainfall of 710 mm.

T76, a 4- to 5-year-old Amazon cultivar, yielded roughly 2200 kg ha-1 of wet beans for each of the three water applications. None of the interventions altered the harvest's seasonality.

A two-year study in Côte d'Ivoire compared two kinds of irrigation (sprinkler and drip) to an unirrigated control treatment on the growth and yield of young cocoa.

When the observed soil water deficit reached 20 mm, sprinkler irrigation was applied, whereas drip irrigation was applied when the soil water tension reached 20 kPa at a depth of 0.20 m.." As a result of these scheduling variations, sprinklers applied 535 mm of water in the dry season compared to drip irrigation's 220 mm (224 mm).

Rose flower count and yields were not affected by drip irrigating although the growth cycle was accelerated and the number of blooms increased.

Based on an unreplicated field irrigation study in Peninsular Malaysia, it is reported that a 0.5-hectare block of mixed hybrid seedlings on a beachfront estate received daily drip irrigation (marine clay).

There was no rain for at least two weeks before irrigation was utilized, unless on days after at least 5 millimeters of rain had fallen or it was actively raining. The unirrigated control was an identical 0.5 ha block. The trial dragged on for over three years.

By irrigation, annual dry bean yields rose from 1500 kilograms per hectare to 2400 kilograms per hectare in the first year, and from 1150 kilograms per hectare to 1450 kilograms per hectare in the second year. In addition, the number of pods grew by an average of 39 percent and the weight of the beans by 7 percent.[3]

The amount of water that was used was not indicated. These results are solely indicative of how this particular site responds to irrigation, and they are of limited generalizability to other locations.

How irrigation could encourage Amelonado trees to bloom during the dry season was investigated. An orchard in Ghana was to be pollinated by hand using pollen from this.

Irrigated trees produced 30% more blooms than non-irrigated ones when cherelles were regularly removed, despite the same flowering patterns in both cases. Over-tree sprinklers (used to relieve internal water stress by boosting humidity) were ineffective compared to micro-sprinklers placed under the trees.[4]

Tree survival

For three years, researchers presented novel data on how cacao trees that were given diverse treatments survived.[5]

The researchers found that towards the end of the second year, 74%, 86%, and 82.2% of the trees survived under the light, medium, and deep shade treatments, respectively. By the conclusion of the second year, there were significant differences (p 0.001) in the survival rates of the cacao between the enforced treatments of shade and mulch/irrigation.

There were 70%, 82%, 91% and 94% tree survival rates for each of the soil moisture management treatments (p 0.022).

It was also found that some clones fared worse under no mulch than others. At the conclusion of the second year, the percentage of trees that survived without mulch and watering was 80 and 98 percent, respectively.[5]

Tree Yield

According to the researchers, there were significant differences in dry cocoa bean production between genotypes in the first year of cropping (p = 0.011). Mulch/irrigation treatments had significant yield differences (p = 0.001), with irrigation and coffee husk treatments producing the highest yields, and plastic mulch and zero mulch producing the lowest yields.

Growth of trees

There was a noticeable increase in tree stem volume in the treated areas, which was likely due to the increased photosynthesis during the dry season. Cacao's photosynthetic mechanism can be permanently disrupted by water stress of sufficient intensity.[6]

When trees are mulched with plastic in the dry season, their survival rates can be comparable to that of irrigation.

For example, the mulch treatments also alleviate stress by lowering the root zone temperature (independently of air temperature), which has been shown to affect the survival, growth and dry matter partitioning of field crops.[7]

Trees with mulch have less leaf abscission during the dry season. Researchers have found that preserving soil moisture impacts leaf area duration in cocoa plants.[8]

When trees are watered or have appropriate moisture, the use of coffee husk mulch results in an early and higher yield than the use of plastic mulch. Mulching the soil with coffee husks provides nutrients and helps to conserve water.

Besides coffee husks, other similar organic mulches, such as Weeding waste, decaying plantain pseudostems, falling litter like dried cacao leaves, and cocoa bean shells where pod illnesses aren't an issue, can be used in places where they're not readily available.[9]

Use of Grafted trees

Acheampong and colleagues found that substantial losses are incurred when no mulching and light shade is supplied.[5]

When mulching and shading are used in conjunction, trees have a better chance of surviving the dry season.[10]

Source: https://www.foto-grafo.de/Cocoa/cocoa_4.html

A solid shade management strategy can help expand and increase productivity. If you want faster plant development and less disease and pests, minimize the amount of shading during wet seasons.

NOTES

Chapter 1

1. Sepal - Definition and Function | Biology Dictionary. https://biologydictionary.net/sepal/

2. Petal - Wikipedia. https://en.wikipedia.org/wiki/Petal

3. Sphenostylis angustifolia | PlantZAfrica. http://pza.sanbi.org/sphenostylis-angustifolia

Chapter 2

1. Edwards, D. F. (1973). Crosses with the putative T. cacao x Herrania spp. hybridAnnual report 1971-1972. Cocoa Research Institute, Tafo (Ghana).

2. Allison, H. W. S., & Smith, R. W. (1964). Economics of cacao establishment on clear-felledland. Reprinted from World Crops.

3. Gomes, A. S., & Kozlowski, T. T. (1987). Effects of temperature on growth and water relations of cacao (Theobroma cacao var. Comum) seedlings. Plant and soil, 103(1), 3-11.

4. Alvim, P. D. T. (1977). Cacao. Ecophysiology of tropical crops, 279313.

Carr, M. K. V., & Lockwood, G. (2011). The water relations and irrigation requirements of cocoa (Theobroma cacao L.): a review. Experimental agriculture, 47(4), 653-676.

5. McKelvie, A. D. (1956). Cherelle wilt of cacao: I. Pod development and its relation to wilt. Journal of Experimental Botany, 7(2), 252-263.

6. Almeida, A. A. F. D., & Valle, R. R. (2007). Ecophysiology of the cacao tree. Brazilian Journal of Plant Physiology, 19, 425-448.

7. Daymond, A. J., & Hadley, P. (2008). Differential effects of temperature on fruit development and bean quality of contrasting genotypes of cacao (Theobroma cacao). Annals of Applied Biology, 153(2), 175-185.

8. Pang, J. T. Y., & Lockwood, G. (2008). A re-interpretation of hybrid vigour in cocoa. Experimental Agriculture, 44(3), 329-338.

9. Daymond, A. J., Hadley, P., Machado, R. C., & Ng, E. (2002). Genetic variability in partitioning to the yield component of cacao (Theobroma cacao L.). HortScience, 37(5), 799-801.

Chapter 3

1. Knight, R., & Rogers, H. (1955). Incompatibility in Theobroma cacao. Heredity, 9(1), 69-77.

2. Aneja, M., Gianfagna, T., Ng, E., & Badilla, I. (1994). Carbon dioxide treatment partially overcomes self-incompatibility in a cacao genotype. HortScience, 29(1), 15-17.

3. Warren, J., & Misir, S. (1995). Isozyme markers for self-compatibility and yield in Theobroma cacao (cacao). Heredity, 74(4), 354-356.

Chapter 4

1. M. K. Adjaloo, W. Oduro, and B. K. Banful. Floral Phenology of Upper Amazon Cocoa Trees: Implications for Reproduction and Productivity of Cocoa.

https://doi.org/10.5402/2012/461674

2. Stephen F. Omondi , David W. Odee, George O. Ongamo, James I. Kanya, and Damase P. Khasa (2016). Synchrony in Leafing, Flowering, and Fruiting Phenology of Senegalia senegal within Lake Baringo Woodland, Kenya: Implication for Conservation and Tree Improvement.

https://www.hindawi.com/journals/ijfr/2016/690 4834/

Chapter 6

1. Blaser, W. J., Oppong, J., Yeboah, E., & Six, J. (2017). Shade trees have limited benefits for soil fertility in cocoa agroforests. Agriculture, Ecosystems & Environment, 243, 83-91.

2. Hosseini-Bai, S., Trueman, S. J., Nevenimo, T., Hannet, G., Randall, B., & Wallace, H. M. (2019). The effects of tree spacing regime and tree species composition on mineral nutrient composition of cocoa beans and canarium nuts in 8-year-old cocoa plantations. Environmental Science and Pollution Research, 26(21), 22021-22029.

Chapter 7

1.https://4rplus.org/nutrient-management/

2.https://agrilifeextension.tamu.edu/ library/gardening/fertilizing/

Chapter 8

1. http://apps.worldagroforestry.org/treesandmark ets/inaforesta/documents/agrof_cons_biodiv/co coa%20frm%20bud%20to%20bean.htm

2. Mel'nichenko, A. N. (1952). 122 Insect Pollination Of Cultivated Crop Plants. Insect Pollination Of Cultivated Crop Plants, 34, 122.

3. Cortez, C. J. (2009). Pollination Efficiency in Major Trunk and Branch Axis of the Cauliflorous Theobroma cacao L.

4. Gnanaratnam, J. K. (1954). Pollination mechanism of the cacao flower. Trop. Agric., 31, 98-104.

5. Saunders, J. L., & Knoke, J. K. (1967). Diurnal emergence of Xyleborus ferrugineus (Coleoptera: Scolytidae) from cacao trunks in Ecuador and Costa Rica. Annals of the Entomological Society of America, 60(5), 1094-1096.

6. N'Goran, J. A., Laurent, V., Risterucci, A. M., & Lanaud, C. (1994). Comparative genetic diversity studies of Theobroma cacao L. using RFLP and RAPD markers. Heredity, 73(6), 589-597.

Chapter 9

1.
https://extension.entm.purdue.edu/publichealth/insects/bitingmidge.html

2. Bridgemohan, P., Singh, K., Cazoe, E., Perry, G., Mohamed, A., & Bridgemohan, R. S. (2017). Cocoa floral phenology and pollin ation: Implications for productivity in Caribbean Islands. Journal of Plant Breeding and Crop Science, 9(7), 106-117.

Chapter 12

1. COCOA   MANUALhttps://www.worldcocoafoundation.org › vos2003PDF

2. Brooks, E. R., & Guard, A. T. (1952). Vegetative anatomy of Theobroma cacao. Botanical Gazette, 113(4), 444-454.

3. Aneja, M., Gianfagna, T., & Ng, E. (1999). The roles of abscisic acid and ethylene in the abscission and senescence of cocoa flowers. Plant Growth Regulation, 27(3), 149-155.

4. Hasenstein, K. H., & Zavada, M. S. (2001). Auxin modification of the incompatibility response in Theobroma cacao. Physiologia Plantarum, 112(1), 113-118.

5. Thrower, L. B. (1960). Observations on the diseases of Cacao pods in Papua-New Guinea.-I. Fungi associated with mature pods.-II. Cherelle wilt. Tropical Agriculture, 37(2), 111-124.

6. Nichols, R. (1961). Xyiem Occlusions in the Fruit of Cacao (Theobroma cacao) and their Relation to Cherelle Wilt. Annals of Botany, 463-475.

7. Nichols, R. (1964). Studies of fruit development of cacao (Theobroma cacao) in relation to Cherelle Wilt: I. Development of the pericarp. Annals of Botany, 28(4), 619-635.

8. McKelvie, A. D. (1956). Cherelle wilt of cacao: I. Pod development and its relation to wilt. Journal of Experimental Botany, 7(2), 252-263.

9. Valle, R. R., De Almeida, A. A., & De O. Leite, R. M. (1990). Energy costs of flowering, fruiting, and cherelle wilt in cacao. Tree physiology, 6(3), 329-336.

10. Cope, F. W. (1962). The mechanism of pollen incompatibility in Theobroma cacao L. Heredity, 17(2), 157-182.

11. Hasenstein, K. H., & Zavada, M. S. (2001). Auxin modification of the incompatibility response in Theobroma cacao. Physiologia Plantarum, 112(1), 113-118.

12. Baker, R. P., Hasenstein, K. H., & Zavada, M. S. (1997). Hormonal changes after compatible and incompatible pollination in

Theobroma cacao L. HortScience, 32(7), 1231-1234.

13. N'Goran, J. A., Laurent, V., Risterucci, A. M., & Lanaud, C. (1994). Comparative genetic diversity studies of Theobroma cacao L. using RFLP and RAPD markers. Heredity, 73(6), 589-597.

14. Bos, M. M., Steffan-Dewenter, I., & Tscharntke, T. (2007). The contribution of cacao agroforests to the conservation of lower canopy ant and beetle diversity in Indonesia. Biodiversity and Conservation, 16(8), 2429-2444.

15. Falque, M., Lesdalons, C., & Eskes, A. B. (1996). Comparison of two cacao (Theobroma cacao L.) clones for the effect of pollination intensity on fruit set and seed content. Sexual Plant Reproduction, 9(4), 221-227.

16. Alvim, P. D. T. (1954). Studies on the cause of cherelle wilt of cacao. Turrialba (IICA) v. 4 (2) p. 72-78.

17. Humphries, E. C. (1944). A consideration of the factors controlling the opening of buds in the cacao tree (Theobroma cacao). Annals of Botany, 8(30/31), 259-267.

18. Uthaiah, B. C., & Sulladmath, U. V. (1981). Effect of growth regulators on cherelle wilt in cacao, Theobroma cacao L. Journal of Plantation Crops (India) v. 9 (1) p. 46-50.

19. Orchard, J., Collin, H. A., Hardwick, K., & ISAAC, S. (1994). Changes in morphology and measurement of cytokinin levels during the development of witches' brooms on cocoa. Plant Pathology, 43(1), 65-72.

20. Santoso, D., & Purwanto, R. (2013). Chlorocholine chloride induces cacao reproductive development leading to improved fruitlets productivity of cacao trees in the field. Journal of Agricultural Science and Technology. B, 3(7B), 517.

21. Naundorf, G. (1951). La producción química de nuevas variedades de plantas. Acta Agronómica, 1(1), 15-26.

22. Ahenkorah, Y., Halm, B. J., Appiah, M. R., Akrofi, G. S., & Yirenkyi, J. E. K. (1987). Twenty years' results from a shade and fertilizer trial on Amazon cocoa (Theobroma cacao) in Ghana. Experimental agriculture, 23(1), 31-39.

23. Baligar, V. C., & Fageria, N. K. (2005). Soil aluminum effects on growth and nutrition of cacao. Soil Science & Plant Nutrition, 51(5), 709-713.

Chapter 13

1. Adomako, B. (2007). Causes and extent of yield losses in cocoa progenies. Tropical Science, 47(1), 22-25.

2. Lee, G. R. (1975). Irrigated Upper Amazon cacao in the Lower Shire Valley of Malawi, 1: NPK factorial trial. Tropical Agriculture (Trinidad y Tobago) v. 52 (1) p. 65-69.

3. Carr, M. K. V., & Lockwood, G. (2011). The water relations and irrigation requirements of cocoa (Theobroma cacao L.): a review. Experimental agriculture, 47(4), 653-676.

4. Joly, R. J., & Hahn, D. T. (1989). Net CO_2 assimilation of cacao seedlings during periods of plant water deficit. Photosynthesis research, 21(3), 151-159.

5. Acheampong, E. O., Sayer, J., & Macgregor, C. J. (2018). Road improvement enhances smallholder productivity and reduces forest encroachment in Ghana. Environmental Science & Policy, 85, 64-71.

6. Joly, R. J., & Hahn, D. T. (1989). Net CO_2 assimilation of cacao seedlings during periods of plant water deficit. Photosynthesis research, 21(3), 151-159.

7. de Abreu, J. M., Nakayama, K., Benton, F. P., da Cruz, P. F. N., Ferraz, E. C. A., Menezes, M., & Smith F, G. E. (1989). Cacao pest control in Bahia, Brazil. Cacao pest control in Bahia, Brazil.

8. Joly, R. J., & Hahn, D. T. (1989). Net CO 2 assimilation of cacao seedlings during periods of plant water deficit. Photosynthesis research, 21(3), 151-159.

9. Teye, E., Huang, X., Sam-Amoah, L. K., Takrama, J., Boison, D., Botchway, F., & Kumi, F. (2015). Estimating cocoa bean parameters by FT-NIRS and chemometrics analysis. Food chemistry, 176, 403-410.

10. Jadin, P., & Jaquemart, J. P. (1978). The effect of irrigation on the earliness of young cocoa trees. Cafe Cacao The.

BIBLIOGRAPHY

A. M. Young, Population Biology of Tropical Insects, Plenum Press, New York, NY, USA, 1982.

A.M. Young, "Habitat differences in cocoa tree flowering, fruitset, and pollinator availability in Costa Rica," Journal of Tropical Ecology, vol. 2, no. 2, pp. 163–186, 1986.

E. A. Frimpong, I. Gordon, P. K. Kwapong, and B. Gemmill-Herren, "Dynamics of cocoa pollination: tools and applications for surveying and monitoring cocoa pollinators," International Journal of Tropical Insect Science, vol. 29, no. 2, pp. 62–69, 2009.

D. R. Glendinning, "Natural pollination of Cocoa," New Phytologist, vol. 71, no. 4, pp. 719–729, 1972.

https://www.cocobod.gh/home_section.php?sec=1

http://www.fao.org/3/ad220e/AD220E04.htm

https://www.myagricworld.com/2018/01/04/when-to-apply-fertilizer-on-a-cocoa-farm/

Plantation Crops :: Cocoa - Horticulture

agritech.tnau.ac.in › horticulture ›
horti_plantation cro..

The shade and fertiliser requirements of cacao
(<i ...

onlinelibrary.wiley.com › doi › jsfa.2740130401
› pdf

Growing Cacao – Cacao Production Guide

businessdiary.com.ph › AgriBusiness

nutrient management strategy required for the
soils of the ...

www.icco.org › about-us › doc_download ›
3479-usin...

The soil diagnosis method to compute cocoa
fertilizer ...

agritrop.cirad.fr › document_554340

knepublishing.com › KnE-Life › article › view

www.intechopen.com › books › challenges-in-cocoa-p...

www.sciencefriday.com › articles › meet-the-flies-that-...

www.sciencenewsforstudents.org › article › flowers-cho...

nph.onlinelibrary.wiley.com › j.1469-8137.1972.tb01284.x

www.hindawi.com › journals

www.tandfonline.com › doi › pdf